WORLD'S WORST DISASTERS

FIRES
THE WORST IN HISTORY

BY JENNA VALE

Gareth Stevens
PUBLISHING

Please visit our website, www.garethstevens.com. For a free color catalog of all our high-quality books, call toll free 1-800-542-2595 or fax 1-877-542-2596.

Library of Congress Cataloging-in-Publication Data

Names: Vale, Jenna, author.
Title: Fires : the worst in history / Jenna Vale.
Description: New York : Gareth Stevens Publishing, [2025] | Series: World's worst disasters | Includes bibliographical references and index.
Identifiers: LCCN 2024001353 | ISBN 9781482466645 (library binding) | ISBN 9781482466638 (paperback) | ISBN 9781482466652 (ebook)
Subjects: LCSH: Fires–History–Juvenile literature. | Wildfires–History–Juvenile literature. | Fire prevention–Juvenile literature.
Classification: LCC TH9448 .V35 2025 | DDC 363.3709–dc23/eng/20240216
LC record available at https://lccn.loc.gov/2024001353

First Edition

Published in 2025 by
Gareth Stevens Publishing
2544 Clinton St
Buffalo, NY 14224

Portions of this work were originally authored by Janey Levy and published as *World's Worst Fires.* All new material in this edition is authored by Jenna Vale.

Designer: Claire Zimmermann
Editor: Megan Kellerman

Photo credits: Cover, p. 1 (main photo) anthony heflin/Shutterstock.com; series art (newsprint background) Here/Shutterstock.com; series art (fact box background) levan828/Shutterstock.com; series art (cover & caption box paper texture) Suti Stock Photo/Shutterstock.com; series art (hazard symbol) Maksym Drozd/Shutterstock.com; p. 5 JP Phillippe/Shutterstock.com; p. 7 Steve Sanchez Photos/Shutterstock.com; p. 9 1871 Great Chicago Fire destroyed buildings.jpg/Wikimedia Commons; p. 11 (graphic) Macrovector/Shutterstock.com; p. 11 (background) FJAH/Shutterstock.com; p. 12 Wildnerdpix/Shutterstock.com; p. 13 Marius Dobilas/Shutterstock.com; p. 17 Afrianto Silalahi/Shutterstock.com; p. 19 (upper) ZUMA Press, Inc./Alamy Stock Photo; p. 19 (lower) M Yerman/Shutterstock.com; p. 21 Abaca Press/Alamy Stock Photo; p. 23 David J. Mitchell/Shutterstock.com; p. 25 (upper) Master Sgt. Andrew Jackson/Defense Visual nformation Service; p. 25 (lower) Dominick Del Vecchio/Defense Visual Information Service; p. 27 Andrey_Popov/Shutterstock.com; p. 29 Henrik Dolle/Shutterstock.com.

Printed in the United States of America

CPSIA compliance information: Batch #CS25GS: For further information contact Gareth Stevens at 1-800-542-2595.

CONTENTS

WORDS IN THE GLOSSARY APPEAR IN **BOLD** THE FIRST TIME THEY ARE USED IN THE TEXT.

FORCE OF NATURE

Fire is a natural force that has many uses. It cooks our food and warms our homes. Candles provide light when the power goes out, and building a campfire can be fun. Small, controlled fires have helped people in their daily lives for hundreds of thousands of years.

When fire gets out of control, however, it can cause a lot of damage, or harm. If fire has a lot of **fuel**, it can be hard to stop it from spreading. This can happen in towns, cities, or out in nature. Fires that burn in forests and fields are called wildfires. Without safe conditions, fire can be very dangerous.

AUSTRALIA'S BLACK SATURDAY BUSHFIRES BURNED FOR OVER A MONTH IN 2009. FALLEN POWER LINES WERE ONE MAJOR CAUSE.

BLACK SATURDAY BUSHFIRES

WHEN A FIRE DESTROYS A LARGE AREA AND HURTS OR KILLS PEOPLE, IT IS CALLED A NATURAL DISASTER. IN 2009, SUCH A DISASTER HIT QUEENSLAND, AUSTRALIA. AS MANY AS 400 INDIVIDUAL BUSHFIRES BURNED AROUND 1,737 SQUARE MILES (4,500 SQ KM) OF LAND. MORE THAN 2,000 HOMES WERE DESTROYED, AND 173 PEOPLE DIED. IT WAS ONE OF THE WORST NATURAL DISASTERS IN AUSTRALIA'S HISTORY.

IN THE CITY

People should always pay attention when using fire or anything that could cause a fire. Old or broken household appliances or electrical outlets can cause fires. Grills, gas stoves, and **arson** can cause fires too.

In cities, fire and smoke can spread quickly inside or between big buildings that are close together. Wood, tar, and other building matter that are common in cities supply fuel for a fire. When a city fire is finished burning, many people may be homeless and need help finding somewhere else to live. The damage from a city fire may take a long time to repair.

A SPACE HEATER CAUSED A FIRE IN THIS BRONX APARTMENT BUILDING IN 2022. IT TOOK AROUND 200 FIREFIGHTERS TWO HOURS TO PUT OUT THE FLAMES.

A DEADLY SPARK

IN JANUARY 2022, AN APARTMENT BUILDING IN THE BRONX, NEW YORK, CAUGHT FIRE. A **FAULTY** SPACE HEATER CAUSED THE BLAZE. AN OPEN DOOR IN THE APARTMENT ALLOWED SMOKE TO SPREAD THROUGHOUT THE BUILDING, FILLING THE HALLWAYS AND STAIRWELLS. SEVENTEEN PEOPLE DIED, AND DOZENS MORE WERE HURT. IT WAS THE DEADLIEST FIRE IN NEW YORK CITY THIS CENTURY.

THE GREAT CHICAGO FIRE

Drought is another condition that allows fire to spread easily. In 1871, the city of Chicago, Illinois, faced a drought for two months. The city's wooden structures became very dry, making them more **flammable**. Many small fires broke out in the city, but one fire became huge in under two hours.

The fire began October 8 in a barn. Strong wind helped the fire spread quickly, and tired firefighters arrived too late to stop it. The fire burned for over 24 hours. It destroyed downtown Chicago and killed more than 300 people. About 100,000 people were left homeless. It was one of the worst city fires in U.S. history.

IN THIS PHOTO, PEOPLE SURVEY SOME OF THE DESTROYED BUILDINGS AFTER THE GREAT CHICAGO FIRE. THE FIRE CAUSED $200 MILLION OF DAMAGE TO THE CITY.

REBUILDING CHICAGO

THE 1871 FIRE DESTROYED AROUND ONE-THIRD OF CHICAGO'S BUILDINGS. WHEN THE CITY BEGAN REBUILDING, THE NEW STRUCTURES WERE SIMILAR TO THOSE THAT HAD BURNED DOWN. IN 1874, THERE WAS ANOTHER BIG FIRE IN CHICAGO. IT WAS ONLY AFTER THE 1874 FIRE THAT THE CITY BEGAN REQUIRING BUILDERS TO USE MATERIALS THAT WERE LESS FLAMMABLE.

OUT IN THE WILD

Wildfires can be particularly dangerous because they can be even harder to manage than city fires. Wildfires are caused by lightning, people not being careful, and arson. Wind and dry weather can make a small fire grow very large and out of control in minutes.

Wildfires burn 5 million to 10 million acres (2 million to 4 million ha) of land yearly around the world. Wildfires kill animals, trees, plants, and can damage the land. A forest can take years or even centuries to grow back. If a wildfire spreads far enough, it can affect towns and cities as well.

WILDFIRE STAGES

DROUGHT MAKES IT EASIER FOR WILDFIRES TO SPREAD. IF THE FIRE GETS HOT ENOUGH, IT CAN EVEN DRY OUT PINE NEEDLES AND LEAVES SO THEY **IGNITE** MORE EASILY.

AFTEREFFECTS

IF A WILDFIRE BURNS TOO HOT AND FOR TOO LONG, IT CAN SEVERELY DAMAGE THE ECOSYSTEM. IT MIGHT BURN SO MUCH LAND THAT IT RUINS THE SOIL, AND ANIMALS LOSE THEIR HABITATS AND FOOD SOURCES. GOVERNMENT FOREST SERVICES THEN HAVE TO EXAMINE THE AREA TO SEE HOW THEY CAN RESTORE THE ECOSYSTEM AND PREVENT FUTURE PROBLEMS.

Firefighters slow wildfires in a few ways. They spray the fire with water or chemicals. They may also dig up the ground to create a firebreak, which is a wide band around the fire. They can also start a small fire, called a backfire, to remove fuel between the firebreak and the wildfire.

Sometimes firefighters need to get behind the lines of a wildfire to make a firebreak. To do this, they fly over the fire in aircraft and jump out with **parachutes**. These special firefighters are called smokejumpers. They bring supplies such as food, water, and even a sleeping bag in case it takes a long time to slow the fire.

SOMETIMES FIREFIGHTERS USE HELICOPTERS OR AIRPLANES TO DROP WATER ON WILDFIRES FROM ABOVE.

CONTROLLED BURNING

AS SCARY AS IT MIGHT SEEM, SOMETIMES IT'S BETTER TO LET WILDFIRES BURN OUT NATURALLY. SOMETIMES GOVERNMENT FOREST SERVICES EVEN SET FIRE TO WOODED AREAS ON PURPOSE. THIS KIND OF FIRE IS CALLED CONTROLLED BURNING, AND IT CAN HELP KEEP THE FOREST HEALTHY BY BURNING AWAY DEAD PLANT MATTER OR **INVASIVE SPECIES** OF INSECTS OR PLANTS.

PESHTIGO IN 1871

On the same day as the Great Chicago Fire in 1871, wildfires raged across the Upper Midwest region of the United States. The worst burned through the town of Peshtigo, Wisconsin. It was even deadlier than the fire in Chicago.

Like Chicago, Wisconsin had a drought in 1871. Small wildfires started. Strong winds spread the fires, and they came together. The wildfire spread so fast that people could not escape. Peshtigo and several other villages were destroyed, and the fire burned more than 1 million acres (405,000 ha). It was the deadliest wildfire in U.S. history, killing between 1,200 and 2,400 people.

DROUGHT
several months of dry weather

WARM TEMPERATURES
a much warmer fall season than usual

LOGGING DEBRIS
dry branches and other debris left behind by Michigan loggers

THE GREAT FIRES OF 1871

fires in Chicago, Michigan, and Wisconsin

FLAMMABLE MATERIALS
wooden structures and buildings

STRONG WINDS
fuel that allowed small fires to grow and spread

⚠ THESE ARE A FEW OF THE CONDITIONS THAT LED TO THE MANY FIRES OF 1871. THE CAUSE OF EACH FIRE IS UNKNOWN, BUT THESE FACTORS TOGETHER MADE THE FIRES **CATASTROPHIC**.

FURTHER BURNING

OTHER BIG FIRES HAPPENED AT THE SAME TIME AS THE GREAT CHICAGO FIRE AND THE PESHTIGO FIRE. WILDFIRES BURNED THROUGH FORESTS IN MUCH OF WISCONSIN AND MICHIGAN. IN MICHIGAN, THREE MAJOR FIRES IN THE CITIES OF PORT HURON, HOLLAND, AND MANISTEE WOULD TOGETHER BECOME KNOWN AS THE GREAT MICHIGAN FIRE.

OUT OF CONTROL

In recent decades, the Southeast Asian country of Indonesia has had several intense fire seasons. The worst happened in 1997. Smoke from the fires even caused problems in nearby nations.

In Indonesia, big companies start fires to clear away trees and plants. These fires got out of control in 1997. The fires spread and burned almost 20 million acres (8 million ha). Smoke from the fires spread as far as Australia. Hundreds of people died, and the smoke made tens of thousands of people sick. Such fires have become an annual problem in Indonesia. Fires in 2015 and 2023 were especially bad, creating a **haze** that spread to nearby countries again.

EL NIÑO

THE INDONESIAN FIRES IN 1997, 2015, AND 2023 OCCURRED DURING EL NIÑO YEARS. EL NIÑO IS AN EVENT THAT CHANGES HOW WATER FLOWS IN THE PACIFIC OCEAN. THIS CHANGE AFFECTS THE WEATHER, INCLUDING WHERE RAIN FALLS. IN INDONESIA'S CASE, EL NIÑO CAUSED A DROUGHT ALL THREE YEARS BECAUSE RAIN DIDN'T FALL WHERE IT NORMALLY WOULD.

HERE, FIREFIGHTERS TRY TO PUT OUT PEATLAND FIRES IN INDONESIA IN 2020. PEATLAND FIRES CAN BE ESPECIALLY DANGEROUS BECAUSE THE DEAD PLANT MATTER THAT MAKES UP PEAT CAN FUEL FIRES FOR MONTHS.

CAMP FIRE

California deals with wildfires every year. In 2018, California fought the Camp Fire, its deadliest wildfire yet. It was called the Camp Fire because it started near Camp Creek Road in Butte County. A faulty electric transmission line caused the fire.

Wind blew the fire downhill toward the town of Paradise. The fire spread quickly and destroyed around 11,000 homes. It killed 85 people and hurt 17 others. Within a few hours, most of Paradise was destroyed. Nearly 5,600 firefighters worked together to fight the flames during the worst of the blaze. In the end, the Camp Fire burned across more than 150,000 acres (60,700 ha).

FIRESTORM

WHEN THE CAMP FIRE BLEW DOWNHILL TOWARD PARADISE, IT FORMED A FIRESTORM. A FIRESTORM IS AN EVENT IN WHICH THE HEAT FROM A WILDFIRE CREATES POWERFUL WINDS. HOT AIR RISES, AND IN A FIRESTORM, A LOT OF HOT AIR RISES VERY QUICKLY. THIS CAUSES AN UPDRAFT, OR WIND MOVING UPWARD, TO REPLACE THE RISING HOT AIR.

THE CAMP FIRE SPREAD TO MANY HOUSES, INCLUDING THE ONE SEEN HERE, IGNITING ENTIRE NEIGHBORHOODS.

REMAINS OF A HOME DESTROYED BY THE CAMP FIRE

MEGAFIRES

Some wildfires become so big that experts call them megafires. These fires burn through 100,000 acres (40,500 ha) or more and greatly affect the environment, people who live in the area, or both. Megafires cause much more damage than smaller, more spread-out fires.

In 2020, several megafires burned during California's wildfire season. Three of the megafires burned along the Sierra Nevada mountain range: the North Complex Fire, the Creek Fire, and the SQF Complex Fire. These were three of the largest fires across the Sierra Nevada in the last 100 years. More than 3,500 homes were destroyed, and much of the land was severely burned.

IN THIS PHOTO, TIRED FIREFIGHTERS TAKE A BREAK FROM BATTLING THE NORTH COMPLEX FIRE IN BERRY CREEK, CALIFORNIA. THE TOWN WAS SEVERELY DAMAGED BY THE MEGAFIRE.

CLIMATE CHANGE

MEGAFIRES HAVE BECOME MORE COMMON IN RECENT YEARS, PARTLY BECAUSE OF **CLIMATE CHANGE**. DRY SEASONS LAST LONGER THAN THEY USED TO, AND THE AIR IS GETTING WARMER. WITH THESE CONDITIONS, SNOW MELTS MORE QUICKLY, AND LAND DRIES OUT FASTER. THE DRIER FORESTS AND FIELDS BECOME, THE MORE LIKELY THEY ARE TO CATCH FIRE.

CANADA IN 2023

Like California, Canada has a wildfire season every year. In 2023, drought and high temperatures made wildfires worse across the country. Megafires formed and caused more damage than any wildfire season in Canada's recorded history. Between June and October, more than 42 million acres (17 million ha) burned.

Thick smoke spread across Canada and even reached many U.S. states. In 11 states, the air became dangerous to breathe on some days while the wildfires burned. Some states had never dealt with such poor air quality. In late June, the smoke reached as far away as Europe.

THIS PHOTO SHOWS HOW SMOKE FROM ONE OF THE MANY 2023 WILDFIRES IN BRITISH COLUMBIA, CANADA, FILLED THE AIR.

TRACKING FIRES

SCIENTISTS KEEP TRACK OF FIRES AND THEIR EFFECTS IN A FEW WAYS. THEY MEASURE CARBON **EMISSIONS** AND FOLLOW CHANGES IN AIR QUALITY. THEY ALSO USE SATELLITE DATA FROM SPACE. GOVERNMENT AGENCIES CAN SEE HOW THE FIRES ARE SPREADING AND WARN PEOPLE TO LEAVE THE AREA IF IT GETS TOO DANGEROUS.

THE MAUI WILDFIRES

In August 2023, the island of Maui, Hawaii, faced the deadliest wildfires the United States had seen in a century. The worst of these burned through the town of Lahaina on the west side of the island. Downed power lines caused the Lahaina fire, which grew quickly because of strong winds.

Around 100 people died and at least 2,200 buildings, mostly homes, were destroyed. In the first few months after the fire, government agents removed **hazardous** materials from damaged or destroyed buildings as part of the cleanup. Families had to wait until it was safe to return to what was left of their homes.

IN THIS PHOTO, SOLDIERS FROM THE HAWAIIAN NATIONAL GUARD CHECK DESTROYED BUILDINGS IN LAHAINA FOR ANYONE WHO MIGHT NEED HELP.

CLEANING UP

LOCAL AND FEDERAL GOVERNMENT AGENCIES WORKED TOGETHER TO CLEAN UP HAZARDOUS MATERIALS AFTER THE FIRES. THEY REMOVED GAS TANKS, CHEMICALS, AND OTHER HARMFUL MATERIALS FROM DESTROYED BUILDINGS. THEY ALSO TESTED ASH FROM THE FIRES AND FOUND THAT IT CONTAINED DANGEROUS HEAVY METALS SUCH AS LEAD AND ARSENIC, WHICH COULD HURT PEOPLE IF BREATHED IN.

SEARCH AND RESCUE TEAMS IN MAUI

KEEPING SAFE AT HOME

While wildfires are more likely to happen in some areas than others, fires can break out any place that provides a spark and fuel. It's important to learn how to prevent a fire or how to escape if one begins.

Make sure your home has working smoke alarms. If you build a fire outside, be sure it is totally out before you leave it. Listen to the people in charge if there is a fire nearby. Your family should also have an escape plan. If a fire starts at home, the goal of your plan should be to get everyone outside safely within two minutes.

TALK TO YOUR PARENT OR GUARDIAN ABOUT TESTING YOUR SMOKE ALARMS REGULARLY. IT'S ALSO A GOOD IDEA TO CHECK FOR ANY FIRE HAZARDS AROUND YOUR HOME.

PREVENT DISASTER

THERE ARE MANY WAYS TO PREVENT A FIRE FROM BREAKING OUT AT HOME. MAKE SURE ANY APPLIANCES THAT GET HOT HAVE PLENTY OF ROOM AROUND THEM. DON'T LEAVE ELECTRONICS PLUGGED IN AND SWITCHED ON WHEN YOU'RE NOT USING THEM. IF YOUR FAMILY LIGHTS CANDLES AT HOME, MAKE SURE THEY'RE BLOWN OUT BEFORE LEAVING HOME OR GOING TO BED.

FIRE SAFETY FOR ALL

Fire safety is important outside your home too. The U.S. Forest Service offers advice on how people can protect themselves and their communities from wildfire damage. In California, power companies have started shutting off electricity when it's very windy to prevent fires from starting when power lines go down.

Because of climate change and other factors, wildfires are happening more often. Even areas that don't have wildfires can be affected if smoke from a big fire reaches them. No matter where you live, learning about fire safety at home and in your community can help you stay safe in the event of a fire.

FIREFIGHTERS GO THROUGH A LOT OF TRAINING TO HELP KEEP PEOPLE SAFE FROM FIRES BIG AND SMALL.

ON THE LOOKOUT

THE U.S. FOREST SERVICE KEEPS LOOKOUT TOWERS IN AREAS WHERE WILDFIRES ARE LIKELY TO HAPPEN. THE TOWERS ARE BUILT ON HIGH GROUND SO THE PERSON ON LOOKOUT DUTY CAN GET A GOOD VIEW OF A LARGE AREA. IF A WILDFIRE STARTS, THE LOOKOUT CAN REPORT IT. THIS HELPS KEEP PEOPLE AND NEARBY COMMUNITIES SAFE.

GLOSSARY

arson: The act of starting a fire on purpose with harmful intent.

catastrophic: Tragic and severely damaging.

climate change: Long-term change in Earth's climate, caused mainly by human activities such as burning oil and natural gas.

drought: A long period of very dry weather.

emissions: Particles that are let out into the air.

faulty: Not working correctly.

flammable: Able to catch fire quickly.

fuel: Something used to make energy, heat, or power.

hazardous: Harmful or dangerous.

haze: Cloudy appearance.

ignite: To catch fire.

invasive species: One kind of living thing likely to spread and be harmful when placed in a new area.

parachutes: Specially shaped pieces of cloth that collect air to slow something down.

FOR MORE INFORMATION

BOOKS

Bard, Breena. *Wildfire.* New York, NY: Little, Brown Ink, 2023.

Hopkinson, Deborah. *The Deadliest Fires Then and Now.* New York, NY: Scholastic, 2022.

Paley, Dan, and Molly Mendoza. *They Hold the Line: Wildfires, Wildlands, and the Firefighters Who Brave Them.* San Francisco, CA: Chronicle Books, 2023.

WEBSITES

National Park Service
nps.gov/subjects/fire/fire-basics-for-kids.htm
Learn about wildfire's role in different ecosystems with articles, photos, and more.

Ready Kids
ready.gov/kids/disaster-facts/wildfires
Learn more about what to do before, during, and after a wildfire.

Safe Kids Worldwide
safekids.org/tip/fire-safety-tips
Learn more about how to prevent, detect, and escape fire in the home.

Publisher's note to educators and parents: Our editors have carefully reviewed these websites to ensure that they are suitable for students. Many websites change frequently, however, and we cannot guarantee that a site's future contents will continue to meet our high standards of quality and educational value. Be advised that students should be closely supervised whenever they access the internet.

INDEX